Made in the USA
Middletown, DE
10 August 2021

45805644R00057